The Color Nature Library
SEA LIFE

By Dr. MAURICE BURTON
and JANE BURTON

Designed by
DAVID GIBBON

Produced by
TED SMART

CRESCENT BOOKS

First published in Great Britain 1978 by
Colour Library International Ltd.,
Designed by David Gibbon. Produced by Ted Smart.
© Text: Maurice Burton/Jane Burton
© Illustrations: Bruce Coleman Ltd.
Colour separations by La Cromolito, Milan, Italy.
Display and Text filmsetting by Focus Photoset, London, England.
Printed and bound by L.E.G.O. Vicenza, Italy.
Published by Crescent Books, a division of Crown Publishers Inc.
All rights reserved.
Library of Congress Catalogue Card No. 77-17747
CRESCENT 1978

INTRODUCTION

It is a common practice among writers on the subject to refer to the bewildering variety of animal life in the sea. This is by no means unjustified. To a large extent it is bewildering because so much of it is unfamiliar, as compared with the animals we meet on land in our everyday life. The creatures are unfamiliar in appearance as well as in the way they are built and in the things they do. Many can only be described as bizarre and, surprisingly, they have an overall beauty.

If marine animals are today unfamiliar, this is nothing compared with the almost total ignorance, even on the part of learned men, which existed until little more than a century ago. In the early part of the nineteenth century naturalists were beginning to show an interest in them, but the real science of the seas was founded in earnest in the 1870s with the famous voyage of the *Challenger*. This was a British warship which was stripped of her guns and furnished with nets and other collecting gear and, with a party of scientists on board, spent three years sailing round the world, collecting specimens and making studies of everything connected with the seas. The ship travelled 69,000 miles in the course of three years. The results of this and of the studies made on the collections brought back were published in long series of large volumes, the famous "Challenger" reports. A major consequence of this was that marine animals, for the first time, could be properly classified, which is the first step to a full understanding.

The pioneering cruise of the *Challenger* was soon followed by others, notably from the United States, Germany, Holland, France and other countries of western Europe. In addition marine Research Stations sprang up on the coasts all over the world. Outstanding among the personalities in this field was Prince Albert I of Monaco who personally financed his research station and his research ships.

In this book we have selected examples of the different groups of marine animals laid out in the order in which they are classified, beginning with some of the lowest in the animal scale, the sponges, and ending with the highest, in this case, the whales.

One of the first peculiarities to be noted of marine fauna is that a considerable number of the animals are sedentary. That is to say, as soon as the larvae settle on the bottom to change into young adults they fix themselves to a stone, rock or seaweed and stay there for the rest of of their lives. This applies especially to sponges, corals, marine bristleworms, oysters and the like. There are others, like the sea-anemones, which are capable of moving about but do so very slowly and very rarely.

Plants are sedentary and for a good reason. They obtain their food from the soil or from sunshine, so they have no need to move about. It is the same with the marine animals that are sedentary, but in this case it is because their food is brought to them. They either feed on small edible particles floating in the sea or in suspension or on small fishes and other small animals that swim to them.

The next oddity, as compared with land animals, is that so many species living in the sea are built on a radial or 'cartwheel' pattern. Examples are jellyfishes and sea-anemones and others like them, such as all the coral polyps and sea mosses, as well as the starfishes and their relatives. A sedentary animal of this type, such as the sea-anemone, benefits from this structure by being able to stretch up or shrink down; it can thus reach food being carried past it at different levels. In addition, the radial arrangement of the sea-anemone's tentacles enable it to catch food coming from any direction.

It is less easy to see what a starfish gains from its radial plan, especially since, when a starfish is going places, one arm takes the lead, so that temporarily the animal can be said to have a front end and a rear end.

The radially-built animals have contributed most to the appearance of what are so often called sea gardens. Their presence on the sea bed invites comparison with flowers, whose petals are often radially arranged and for a similar reason. A sea-anemone needs to have its circlet of tentacles, so reminiscent of a daisy or dahlia-head, because its food comes from all directions. The advantage to the flower of being radially built is that insects, which come and pollinate it, also come from all directions.

Not all sea animals are oddities; some of them are as familiar to us as any animals we know. The numerous fishes are an example, although even there we find some queer fish, like the seahorse, which looks more like a chessman than a living animal. Whales, too, have long been familiar to those who live on the coast or sail the seas. Yet in their own way they are oddities, warmblooded animals that suckle their young yet are shaped like fishes and to a large extent behave like fishes. Indeed, to many seafaring people they are still known as fish. Perhaps the final oddities are the sea cows, the dugong and the manatee, which are said to have given rise to the mermaid story. Christopher Columbus thought they were mermaids, although anything less like the beautiful maidens of the legend is hard to imagine.

Previous page: black-footed penguins
Facing page: bottlenose dolphin

6

Sponges

Very few of the 2500 species of sponge have any commercial value. Most of these are the so-called bath sponges which have fibrous skeletons and are found in warm seas, especially the Mediterranean and the Caribbean. One other kind was used, to a limited extent, as a soil fertiliser. This is the crumb-o'-bread sponge *far left* which is found in many parts of the world, mainly on rocky shores or in shallow offshore seas. It has a skeleton of minute glass (silica) needles.

The crumb-o'-bread sponge, so called because when dried it looks like mouldy bread and readily crumbles, is usually green in life but may be yellow. It shows the typical characteristic of sponges: crater-like openings on the upper surface. All over the body of a sponge are minute pores through which water is drawn, bringing in food (bacteria and microscopic plankton) and oxygen. After the water has circulated through the body it is driven out through the crater-like vents, carrying with it body waste in tiny particles or in solution.

Some sponges have only one large vent instead of several small ones, like the yellow sponge *near left*. It has no common name, although it is very like the sea-fig; this is an irregularly-shaped sponge, usually green but sometimes yellow, which is frequently found growing in the shape of a fig.

Most sponges lack common names and are irregular in shape; they are coloured various shades of cream to yellow, brown, red and purple. Altogether, they look like plants and at one time were thought to be a kind of plant. Good examples are the red sponge of Florida *bottom right* and the brown sponge growing on the Great Barrier Reef *top right;* this seems indistinguishable from the miscellaneous marine growths surrounding it, until we notice its vents.

The sea gardens, as they have been called, of coral reefs owe much of their colour to the various sponges scattered among the corals. In deeper water, however, the colours of sponges and other animals, such as the sea fans growing with the twinned tubular sponges *bottom left* are lost in the general blueness of the underwater scene.

The great majority of sponges live in the seas, being found from the poles to the equator and from mid-tide level on the shore to the greatest ocean depths. There are, however, freshwater sponges in lakes and rivers, green and very plant-like except when growing in the shade, when they are yellow.

7

Jellyfishes

Jellyfishes are marine backboneless animals. They belong to a group of animals known as Cnidaria, or nettle-animals, all of which carry stinging cells. The body of a jellyfish is semi-transparent, umbrella or bell-shaped and jelly-like, sometimes tinted with brown, red, purple, blue or violet as is the violet jellyfish *top left and centre*. The mouth is at the centre of the underside and around it and near the margin, sometimes forming a fringe all round, are the tentacles armed with stinging cells, shown in the illustration of the brown jellyfish *bottom left*. These stinging cells are microscopic but, when a small animal brushes against the surface of a tentacle, they shoot out fine hollow threads that pierce the skin. Poison then flows down the centre of each thread and produces paralysis in the victim.

The food of jellyfishes is usually small fishes and shrimps which swim near the surface. The threads of the nettle-cells not only make them helpless but hold them securely. The more the prey struggles the more firmly it is trapped as it touches still more tentacles which in turn shoot out their poison threads.

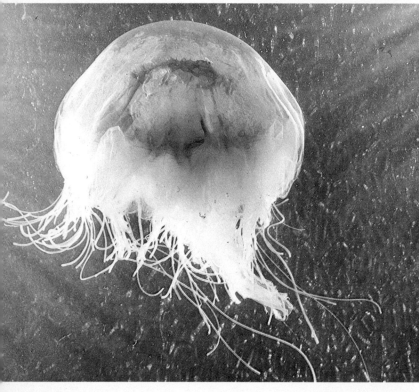

Anemones

These gaily-coloured, almost flower-like, animals are plentiful in rock pools on the seashore and also on the sea bed farther out. They are found from polar seas to the equator but attain their greatest size and magnificence in the warmer seas. Once the larvae have settled, after swimming for a few hours, they grow into small sea-anemones like the beadlet anemones *far right*.

A sea-anemone consists of a central body which is bag-like and has a mouth at the upper end. It is a nettle animal like the jellyfish, with hundreds of stinging cells in its tentacles. This dahlia anemone *top right* is fully expanded, and any small animal swimming near its tentacles is liable to be stung and held. As soon as one tentacle touches something edible the message is conveyed to all the other tentacles which fasten on the prey and convey it to the mouth at the centre.

A sea-anemone normally moves about by creeping on its base. The snakelocks anemone *bottom right*, however, has inflated its body and is walking on its tentacles.

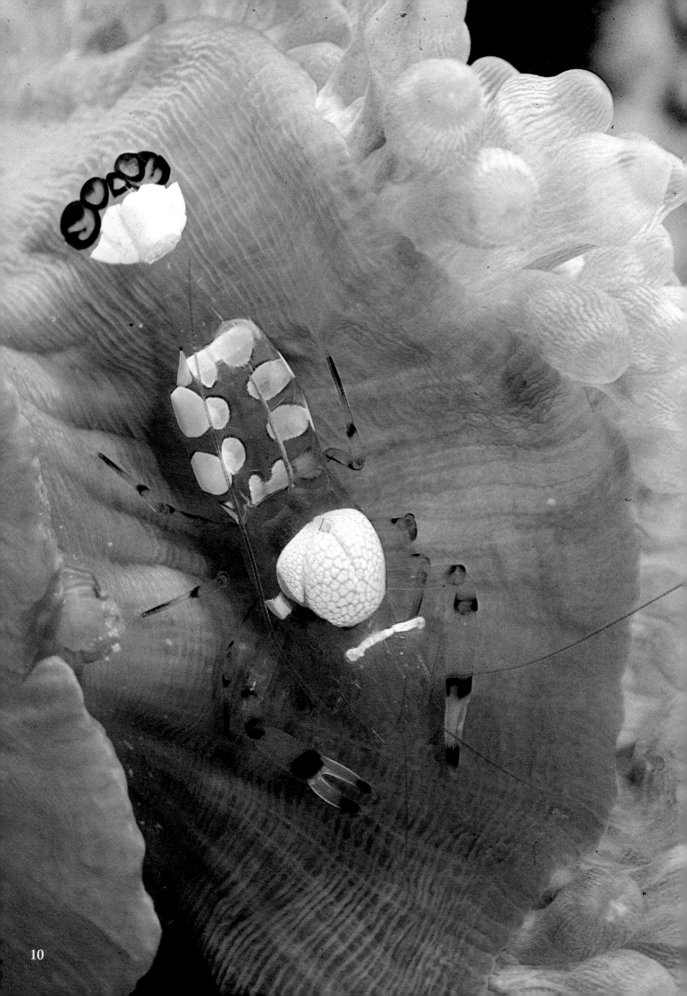

Anemone Partners

Not infrequently two different species live together in very close association. This kind of partnership is known as symbiosis. In some instances the symbiont, or partner in a symbiosis, needs only shelter or protection. This is so with the anemone fish *below*, which is also called a clown fish because of its bright colours. Although the anemone stings small fishes to paralyse them before eating them, the clown fish makes a gradual approach to the anemone which, once it has got used to it, withholds its sting when the clown fish settles on it. The anemone shrimp *left* also shelters against an anemone.

The porcelain crab *right* filters microscopic animals and plants, as well as minute fragments of decaying flesh, from the sea. It is an easy living and the crab makes life even more comfortable by sheltering from possible enemies.

Corals and Sea Fan

There are two kinds of corals: true or stony corals and soft corals. Both are related to sea-anemones but the first has tentacles in multiples of six; the soft corals' are in multiples of eight. The coral animal is called a polyp.

The polyps of stony corals, like the golden tubastraea coral *above*, lay down a hard skeleton of lime on which they sit, so they are unable to move about. Some stony corals consist of a single polyp seated in its stony cup but most are colonial, especially the reef-building corals which include the needle coral *top right* and the staghorn coral *right*.

Soft corals are found in cold as well as warm seas, usually in fairly deep offshore waters but also in the great ocean depths. Most commonly they consist of a stem and branches lying in one plane and are called sea fans. Others, like the purple sea fan *left* branch irregularly but are still called sea fans. The picture *above left* shows polyps of a green coral growing among a purple coralline alga (*Lithothamnion*).

When a coral larva settles on the bottom, changes to a polyp and starts to build its skeleton, a bud appears on its side. At first this is only a small bump.

The bud grows bigger, a mouth appears at the free end and a circle of tentacles grows out around the mouth. When this new bud reaches the size of the parent it also starts to bud. The parent meanwhile has grown another bud. Repeated budding results in a colony that may contain hundreds of thousands of polyps and may be several feet across. No matter how large the coral may be, all the polyps are intimately connected with each other, so that the colony becomes a sheet of animal tissue covering the stony skeleton. On the death of the colony, which in life is usually brightly coloured, the flesh decays leaving only the chalky white skeleton which most people know better as coral.

14

Marine Bristle Worms

Ringed worms include earthworms, marine bristle worms and leeches; they all have the body marked on the surface by rings. The most numerous are the bristle worms and these like earthworms, have setae, or bristles, which may be used for locomotion or for other purposes. The most spectacular are the peacock worms, or fan worms, which have a crown of multi-coloured bristles around the mouth. They and others, with such names as feather-duster worm, live in tubes of mud, slime, sandgrains or chalk, into which they withdraw when disturbed. After a period of quiet the worms protrude again from their tubes *right*, and as they do so the crown of bristles unfolds, to trap food from the water.

Bristle worms contribute in a modest way to the beauty of the undersea scene, as in the colourful conglomerate of marine animals *left*. To appreciate their full beauty, however, they need to be viewed in isolation, as is the fan worm *above*, with its fan of bristles fully spread. This one lives in tropical seas and its delicate beauty loses nothing by comparison with the garish colours of the fish, known as the regal tang, swimming near it.

15

Shrimps and Prawns

Shrimps and prawns are relatives of lobsters and, like them, are protected by a tough coat or crust. They are a few of the many animals making up the group known as crustaceans. Shrimps are usually the smaller but there are other differences. A shrimp has one pair of long antennae and a short pair; a prawn has two pairs of long antennae. The front end of a prawn ends in a sharp beak, as shown in the illustration of the common prawn *below*.

The names 'shrimp' and 'prawn' tend to be used colloquially in a confusing manner. Thus, the mantis prawn *left*, so called because its front legs jackknife like those of the insect, the praying mantis, has the prominent rabbit ears of a typical shrimp. The banded coral shrimp *right* has two pairs of long antennae.

The banded shrimp shown here lost its right claw eleven days before it was photographed. It is growing a new one, the slender transparent limb is seen where the right claw should be.

Lobsters

Lobsters live mainly in shallow off-shore seas hiding in rocky crevices and coming out at night to feed on carrion, which they tear apart with strong claws. When searching for food they walk over the rocks or the sea bed on their five pairs of walking legs. When alarmed they shoot swiftly backwards, swimming through the water like the Pacific lobster *bottom left,* using powerful strokes of the tailfin.

The spiny lobster *right and below,* also known as the painted lobster or marine crayfish, lacks pincers altogether and uses its long, strong antennae to defend itself. A remarkable peculiarity of Caribbean spiny lobsters is that, although normally solitary, they gather together at the end of the summer and perform strange migrations. They form queues *top left,* each lobster grasping

the tail of the one in front. They can be seen marching steadily in parallel lines over the sea bed in the sunlit waters of the Bahamas and off eastern Florida.

Crabs

Crabs are crustaceans which although related to lobsters are very different in appearance and in general habits. The crab's thorax, the main part of the body, is very much broader than a lobster's and the abdomen, which in a lobster forms the tail, is much reduced in size and tucked under the thorax so that it is out of sight until the crab is turned onto its back. The crab's antennae are also much smaller.

Having this broad shape deprives crabs of the streamlining needed for swimming, although some crabs do swim at times, like the velvet swimming crab the front portion of which, shown here *left*, bears the very small antennae and the stalked eyes. In repose, each eye folds down into a groove in the edge of the shell.

The most common crabs are the

shore crabs, two of which can be seen *bottom right*. They have the usual broad flattened shell and the five pairs of appendages making ten limbs in all, whence the group name Decapoda (ten legs). The front pair form strong claws; the other four pairs are walking legs.

Shore crabs vary a great deal in colour. The red individual, the lower of these two, is infested by a parasite, *Sacculina*. This can be seen as a rounded mass on the crab's underside. From this rounded body root-like strands of flesh penetrate the crab's body drawing sustenance from its tissues. The soft-bodied *Sacculina* is, surprisingly, a crustacean also.

Some crabs have exceptionally long legs and are known as spider crabs *top right*. This one, the arrowhead spider crab, from the Caribbean, is eating the partially digested remains of a butterfly fish among the tentacles of a sea-anemone.

Hermit crabs are a class apart. Typically the abdomen, or hind part of the body, is soft, defenceless and slightly spiral. The small swimming legs, seen on the underside of the abdomen of a lobster, are gone except for the last pair. Hermit crabs have taken to living in the empty shells of sea snails. The hermit inspects a shell with its antennae and claws, then, satisfied it is empty and the right size, it turns around and slips its unprotected hind end into the shell, the spiral of the abdomen fitting neatly into the spiral of the shell. As a hermit crab grows in size it needs to transfer to a larger shell. Two hermit crabs are shown, *bottom left* and *below*.

Sea Snails

None of the many kinds of sea snails has played a greater part in human affairs than the cowries. Most cowries live in tropical seas but the European cowrie *top left* is an exception. Another member of the same family is the margin shell *centre*. Cowries have been used as money among undeveloped peoples on oceanic islands; even in parts of India they were used as currency for centuries. They have also had an important place as articles of personal adornment, as natural beads.

Cowries are attractive to shell collectors because of their beautiful colours and, in most species, their glossy surface which looks as if it has been artificially polished. Perhaps the most remarkable feature is that in the living cowrie this beauty is largely hidden by flaps of flesh that reach up over the sides and meet in the middle line on top of the shell.

Early in life a cowrie looks like other sea snails but as growth proceeds the spiral of the shell is enveloped in the final whorl. This can best be understood by looking at the shell of Gray's volute *bottom left* from Western Australia, and the partridge tun *bottom right* of the Great Barrier Reef, in which the final whorl, although large, does not entirely enclose the spiral.

In many species of cowrie the inner whorls become dissolved away. This gives the owner of the shell more room for its body when it withdraws into the protection of the shell. The limy substance of these whorls is absorbed by the mollusk's body and laid down as fresh building material to strengthen the remainder of the shell.

Most sea snails are vegetarians, browsing seaweeds. The common painted topshell *far right* of European seas, here seen on coralline algae, is a browser. Some, by contrast, are active carnivores, like the queen conch, one of the largest of all sea snails. For hunters a well-developed eye is an advantage. This close-up *near right* of the entrance to the conch's shell shows the eye at the centre of the opening.

The shell of the queen conch or of the king conch, which come from the warmer seas of the Gulf of Mexico, Caribbean and Atlantic, can be used as a trumpet by cutting off the tip of the spiral and so making a hole to blow through. Conch shells are exported to Europe, especially Italy, where the shell has long been used for making cameos.

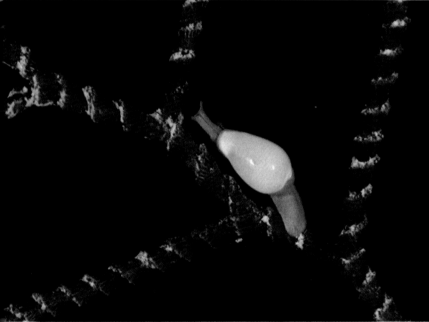

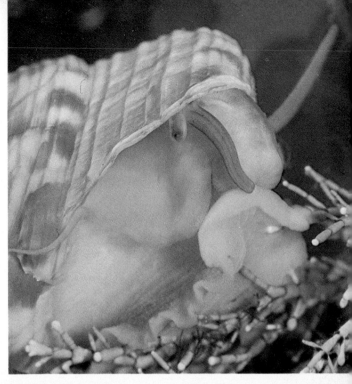

Sea Slugs

Few animals cause such feelings of revulsion as do the slugs, yet they are no more than snails without shells. Sea slugs, by contrast, can be exquisitely beautiful. The body of a land slug is usually seen as a slimy, shapeless mass, with no bright colours to relieve the general drabness.

Sea slugs have much the same form as land slugs but are often very delicately coloured. The body is more flattened than that of a land slug. There is, near the front end, a pair of feelers shaped like rabbit's ears, and near the rear is a bunch of naked gills.

The body of some sea slugs is bedecked on the upper surface by finger-like ceratia, which serve as gills. The European sea slug *left* shows another use for these. It is thrashing its ceratia in a defensive display. The Great Barrier Reef sea slug *below* crawling over a red sea fan, is displaying its gorgeous colours to full effect. On the tropical sea slug *right*, swimming by undulating its body, can be seen the feelers (antennae) and the tuft of reddish gills.

Cephalopods

Sea snails and sea slugs are mollusks. With rare exceptions mollusks are slow-moving, apart from one group which is extremely active. These are the cephalopods, and include the octopuses, squid and cuttlefishes.

Cephalopods have a rounded or torpedo-shaped body with arms beset with suckers surrounding the mouth. The octopuses, here represented *right* by a species that ranges from the north-east Atlantic to the western Pacific, have eight arms. Squid *left* and cuttlefish *below* have two additional arms, much longer than the rest, that are shot out to seize prey, as with a pair of tongs.

Compared with the ordinary run of mollusks, cephalopods are highly organized. The eye is well-developed and so is the brain and the rest of the nervous system. Cephalopods swim backwards projected by a jet of water driven out through a tube on the underside.

Starfishes

Seastars, or starfishes are built on a radial plan, typically with five arms, or arms in multiples of five, radiating from a central body. They are of three kinds, feather stars, starfishes proper and brittlestars or serpent stars.

Feather stars *left* are most nearly related to the sea lilies which live in deep seas and consist of a sort of sea star with feathery arms on a long stalk, the base of which is rooted on the seabed. The feather stars live mainly in shallow seas, have lost the stalk and are mobile. Their feathered arms wave about to catch small edible particles.

Starfishes have a simpler design. The arms are not feathered, the mouth is on the underside and they move about on tube-feet. On the underside of each arm is a groove lined with numerous very small suckers; these are the tube-feet which pull the animal along. They are also used to cling to rocks or to hold, even pull open, the shellfish on which so many starfishes feed. Some starfishes are cushion-shaped *top right*, others have long slender arms *bottom right*.

Brittlestars *below* move with serpentine movements of the arms.

Urchins and Cucumbers

Feather stars, starfishes and brittlestars belong to a branch of the invertebrates known as echinoderms, a name compounded from Greek words meaning spiny skins. There is little to show that they deserve this name, although some of the starfishes are ornamented with blunt spines. We see the reason for the name when we come to close relatives of the seastars, the sea urchins. 'Urchin' was the original name for the hedgehog, the Old World land animal protected by a coat of spines.

Another feature of seastars is that the skin is strengthened by nodules of lime; in the sea urchins these have been converted to plates that fit together to form a complete box, spherical or heart-shaped, inside which lie the animal's essential organs. The box is usually so complete that holes are needed in the plates for communication with the outside world. There are the two normal openings for the digestive system, the mouth being on the underside and furnished with jaws.

As with all echinoderms there is a five-rayed pattern, including five sets of tiny holes running down the sides of the box, through which the tube-feet protrude. There are also five sharp teeth, for browsing seaweeds; these and their supports form a structure which looks like an old-time lantern and is known as Aristotle's lantern, after the Greek philosopher who first studied sea urchins. The spines are set on knobs on the surface of the box, each making a ball-and-socket joint. The sea urchin breathes by gills, small tassels of flesh *top right* that protrude through the limy box to the exterior.

The spines are mainly defensive but some sea urchins walk on their spines, using them like stiff legs. The spines may be short or long and needle-like, and some sea urchins can move their needle-like spines towards any shadow that falls on them. In this way their points are directed at an enemy. The diadematid sea urchins *left* and *centre* do this. In places in the Caribbean these urchins are so crowded on the sea bed that one cannot walk without treading on them, so causing septic wounds.

Sea cucumbers *bottom right* are also echinoderms and the skin, although roughened, is not spiny. They crawl over the sea bed using tube-feet and pick up particles from dead animals with the feathery tentacles surrounding the mouth.

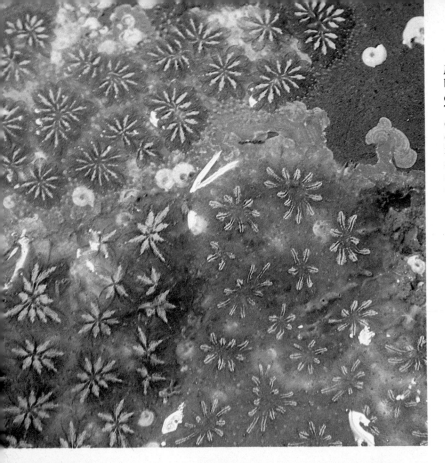

Sea Squirts

Most sea squirts consist of a sac-like body permanently attached to a rock or other solid object, such as pier-piles. There are two openings to the body, one for taking in water, the other for driving it out again. The water brings in minute particles of food and oxygen and carries out body waste. If touched a sea squirt quickly contracts and in doing so squirts out a long thin jet of water. The sea squirt then looks like a lifeless, shapeless bag of jelly, contrasting with the delicate bottle shape *below* or the more common vase shape *right*.

Some sea squirts live in colonies arranged in rosettes or stars, like the golden star sea squirts *left*, all embedded in a common mass of jelly. The individuals forming each rosette share the opening for the exit of the water current.

Sea squirts are believed to be nearly related to the ancestral animals that gave rise to the vertebrates (animals with backbones). This is largely because their tadpole-like larvae have a notochord in the tail. In vertebrates the backbone is first laid down as a notochord in the early stages of the embryo.

Sharks

Most of the world's fishes have bony skeletons. Sharks are a race apart in the fish world. Their skeleton is of cartilage or gristle, not bone. They do not have scales, instead their outer skin is beset with denticles or little teeth which have broad bases and sharp tips and are complete, like any other teeth, with an inner pulp cavity surrounded by dentine; the whole is covered with enamel.

Sharks are primitive. They have been in the seas a very long time, 250 million years. The largest shark living today is the harmless whale shark, which is 17 metres or more long. As might be expected in so primitive an animal, the brain is not highly organized, but sharks make up for this with their agility as swimmers and their ferocious dispositions. Indeed, the mention of their name immediately conjures up a picture of an animal with wide jaws and rows of sharp, vicious teeth.

Carpet sharks *far left* of the western Pacific and Australia, do not conform to the popular idea of sharks. They are flattened from above down and are ornamented with flaps of skin. This, with their colour, means that when they lie motionless on the sea bed, which they habitually do, they look like rocks covered with short growths of seaweed. So instead of pursuing their prey they lie in wait for fishes, crabs and lobsters to wander near enough to be snapped up.

The white tip reef shark *bottom left* of the Indian Ocean and central Pacific, is, in sharp contrast, elegant, torpedo-shaped and built for speed. Any shark, even a small one, will snap at a person when being handled, but those listed as dangerous are the ones known to have attacked human beings in the water.

The common sand shark *right top and bottom*, sometimes called the sand tiger shark, lives in the Mediterranean, Atlantic and South African seas. It is not large, not more than four metres long, but it has been referred to as a wolf of the sea, not just for its voraciousness but more because it hunts in packs. A hundred or more were once seen off the coast of New Jersey herding a shoal of bluefish, themselves ferocious, into a mass in shallow water, and then going into the attack.

Sharks often carry a remora, a fish known as a shark-sucker, clinging by the sucker on top of its head, so getting free transport *top left*.

Rays

The carpet shark, as we have seen, is flattened and lives on the sea bed. The same can be said of rays, which are like sharks in having a skeleton of cartilage. There is, however, one important difference. A shark's gills are on the side of the head, clearly visible because there is no gill-cover as in bony fishes. The gills of rays are on the underside.

Rays usually have a slender tail sharply marked off from the body, as in the blue-spotted reef ray *top left* and the fiddler ray *below*. This is even more pronounced in the Atlantic stingray *bottom left*. Stingrays are so-called because they carry, near the base of the whip-like tail, a saw-edged spine fitted with a poison gland.

The manta ray *right* swims actively, looking like a huge bat.

37

Northern Fishes

A common fish in the North Atlantic, the coalfish *left*, also known as the saithe or coley, is a close relative of the cod which it closely resembles. The cod is a speckled sandy brown on the back with a greenish tinge whereas the coalfish has no speckling, is greenish brown or almost black, with the colour extending down the sides giving a dark appearance. The most obvious difference is that the cod has a pronounced barbel under the chin; the coalfish's barbel is small. Fished commercially, it lives in waters of 100-200 metres.

The bass *top right* is considered one of the best European angling fishes. It reaches a length of a metre and up to 8 kg weight. Although a coastal fish, living in inshore waters, often fished for in the breakers, it tends to go into estuaries or even go up rivers, but goes offshore into deeper waters in winter. Large bass are mainly solitary but can form large shoals at times and in winter are caught in trawls in 6-80 metres of water.

Horse mackerel *centre* have no great economic value, their flesh is not particularly palatable and they do not occur in sufficient numbers, although at times they do come together in shoals. The young are, however, caught in large numbers off Spain and Portugal but are used as fish meal. Also known as scads, jacks, cavallas and pompanos, horse mackerel are not true mackerel. They are, however, well streamlined and usually fast-swimming. The body has a line of small keeled scutes along part of or the entire length of the flank. Young horse mackerel are well known for their habit of sheltering under the bells of jellyfishes, presumably gaining protection by reason of the jellyfish's stinging cells.

The thick-lipped grey mullet *bottom right* is blowing out a mouthful of sand after having extracted anything edible from among it. The food of this, the commonest of mullets, is mainly single-celled microscopic plants, but also includes larval mollusks and crustaceans. This mullet is a common inshore fish of the northern European coasts except in winter when it is believed to hibernate in deeper water taking no food. Since it takes only small food it is notoriously difficult to catch using bait on a line and rod. It is, however, caught in nets locally, but is not economically important.

Reef Fishes

Coral reefs are often referred to as sea gardens because of their massed colour and beauty. The beauty of corals is static, however, but animation is supplied by the many varieties of small fishes that make them their home. Seahorses *left* are not strictly reef fishes yet can often be seen with their tails looped around seaweeds or sea fans associated with reefs. Cardinal fishes *right* are true reef fishes, usually seen in small schools that swim leisurely around the reef, darting back into the recesses of the corals when disturbed. Despite their name they are often brownish although some species are red. Male cardinal fishes fertilize the females internally, which is unusual in fishes, and when the female spawns the male takes the eggs into his mouth and keeps them there until they hatch.

Butterfly-fishes and angel-fishes, *bottom right* are small and highly coloured and are found chiefly on coral reefs. They are deep-bodied and very agile, and they seem to flutter around the coral heads diving into cracks at the first sign of danger. Many browse small seaweeds, others eat the coral polyps or any other small animals living on the reef.

Tropical Fishes

The warmer the seas the greater the expectation of finding gaily-coloured fishes. This is true, however, only for the shallow seas, in which all the bottom-living animals, such as sea anemones and sponges, tend to be brightly coloured. There is an obvious advantage to the fishes in being coloured, since this serves as camouflage. The batfish *left* is so named because the fins recall the wings of a bat. Typical of inshore tropical fishes are the blue angelfish *right* and the yellow-faced butterflyfish *far right* with their striking colours.

In the open ocean tropical fishes are more normally coloured. Whereas the inshore or shallow seas fishes gain a camouflage value from being brightly-coloured, the need of fishes living farther out to sea is to blend as well as possible with the sea and sky, as does the barracuda *below*, the fish that is second only to the sharks in its reputation for ferocity towards humans.

Typical of tropical seas are the flying fishes, like the four-winged flying fish *bottom right* of the Red Sea.

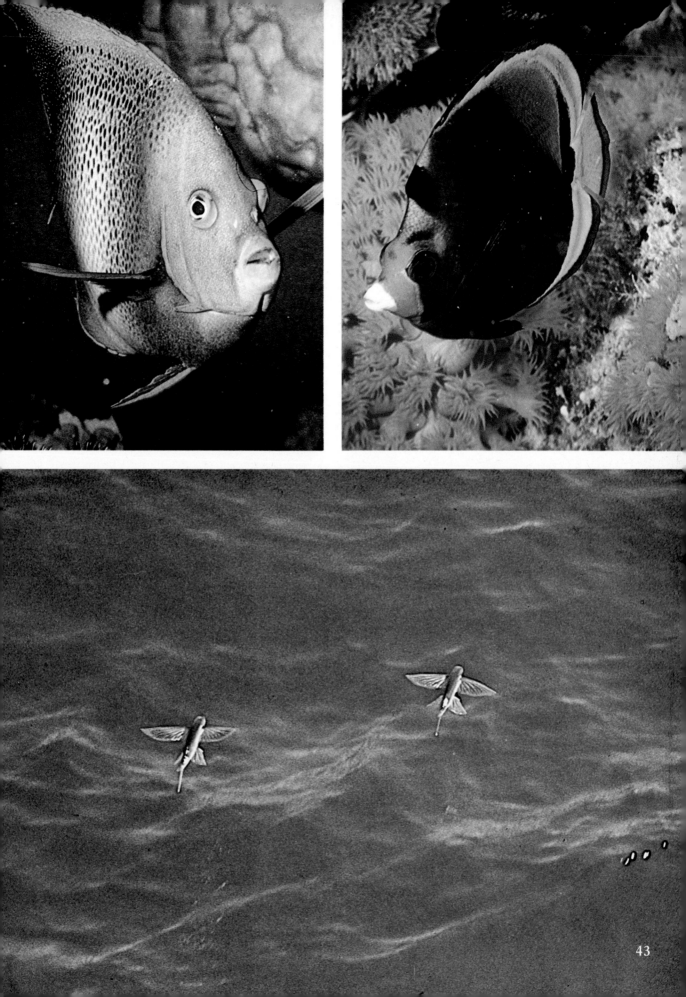

44

Deep Sea

The deep seas are usually reckoned to be waters below a depth of 220 metres. Two-thirds of the surface of the Earth is covered with deep seas. There no light penetrates. Animals living there must either feed on the dead plants and animals (or fragments of dead animals) that continually rain down from the water layers above, or else they must feed on other deep-sea animals.

Since the deep sea is perpetually in darkness, colours are useless, camouflage would be pointless, and we find deep-sea animals are, for the most part, black or sometimes red. The deep-sea prawn *bottom right* is red. Because there is little food there are few animals and these are well spread out. Those, mainly fishes, that feed on other animals must make certain of holding their prey. Large mouths and long, needle-sharp teeth are the rule, like those of the viperfish *top right*.

Perhaps the one outstanding feature of deep-sea animals is their use of light-organs. Deep-sea fishes have rows of lights along the body like the portholes of a liner at night; the pattern of the lights is used to recognise friend or foe or to bring the sexes together. Deep-sea anglerfishes, like *Stomias, centre,* have a luminescent lure to bring prey within snapping distance.

Sargassum Weed

There are two species of seaweed known as Sargassum weed that float freely in the Atlantic in large quantities. The weed is most dense in the area known as the Sargasso Sea and it keeps afloat because of the many small round bladders on its fronds. The seaweed itself is peculiar and has become the home of a number of peculiar animals. One of these animals is the pelagic crab *top left,* a very small crab, widely distributed in tropical and temperate waters and abundant on the floating weed. In fact, the crab will take a ride on anything that floats. Another peculiar animal is the sargassumfish *bottom left,* one of the anglerfishes. It lives permanently on the weed, luring small animals to it with the rod-and-line on top of its head and then snapping them up. If removed from the weed and allowed to swim, it scuttles back rapidly to the shelter of the weed, because that is where it feels safe. It is beautifully camouflaged, being coloured like the sargassum weed amongst which it moves slowly, using its pectoral (breast) fins as arms. To complete the camouflage the outline of the fish is broken up by small irregular flaps of skin.

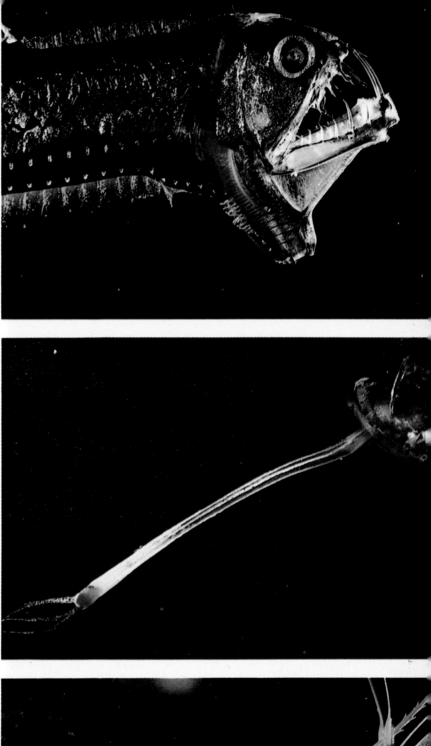

45

Reptiles at Sea

So far as we can make out from the study of fossils and of living animals, life began in the sea and remained there for thousands of millions of years. In due course there began an invasion of land, first by plants and then by animals. When we divide the animals that make up the whole animal kingdom into their natural divisions we find there are twenty-two main groups. All but one of these groups consist of animals without backbones, the invertebrates as they are called. The remaining one consists of vertebrates or animals with backbones.

With only one exception, all the groups of invertebrates are still found exclusively in the sea or have invaded freshwater only; at most they are much less numerous on land than in the sea. This one exception contains a subdivision, the insects, which are overwhelmingly land-living; only a few species have returned to the sea.

The picture for the vertebrates is in striking contrast. Of its five subdivisions, only the fishes are more numerous in the sea than on land, although these have plentifully colonized the lakes and rivers. A few have just managed to obtain a toehold on land, like the mudskippers that daily crawl out on land and stay there for a few hours. The rest of the vertebrates, the amphibians, reptiles, birds and mammals, are typically land-dwelling. Even among these there are a few species that have returned to the sea. These are the species we shall be considering in the rest of this book; the first of them are the reptiles.

The reptiles first came into being about 300 million years ago, as air-breathing land animals, but a few soon took up residence in rivers and lakes, spending most of their time in water, like the crocodiles; a few, such as the now extinct ichthyosaurs and plesiosaurs, became wholly sea-dwelling. Today only the marine turtles and a number of sea snakes, such as the banded sea snake *centre* and *bottom left*, live wholly in the sea.

Sea snakes are found only in the Indian Ocean and in adjacent parts of the Pacific. Any snake can swim if thrown into water, merely by wriggling its body. Sea snakes swim this way and it is made easier for them because the rear end of the body is flattened, like the blade of an oar.

The only other reptile to enter the sea habitually is the marine iguana of the Galapagos *top left and right* and it does so only to feed on seaweed.

46

Marine Turtles

Across the tropical belt the world is spangled with numerous islands large and small, especially in the Caribbean and South Pacific. Usually these islands have sandy beaches. There are also sandy beaches on much of the tropical mainland. Sandy beaches are where turtles lay their eggs, for although these reptiles spend all their time in the sea, they must come on land for this purpose. A Ridley turtle is pictured *below* arriving on a beach in Costa Rica. The female turtle pulls herself up the beach and digs a deep hole in which she lays her 90-100 eggs; she then fills in the hole with sand, using her four flippers, and returns to the sea.

People have been eating turtle eggs for centuries. They have also eaten turtle flesh. The hawksbill turtle has been used for its tortoiseshell, the strong horny layer that covers the bony box which encloses the turtle's body; a hatchling is shown *right* and a yearling *left*.

Except when females come on land to lay, turtles are not often seen. We know they feed on seaweed and a plant known as sea grass, chewing it with their horny parrot-like beaks–they have no teeth.

Sea Birds

Longshoremen work on ships in port or along the shore; mariners put to sea. Most sea birds are longshoremen. The nearest to ocean-going mariners are the albatrosses, like the light-mantled sooty albatross *left*. Albatrosses only come to land to breed. For the rest of the time they fly over the oceans, especially in the windiest parts, like the Roaring Forties.

The masked booby *far left,* one of the gannet family, is a large sea bird, stream-lined for flight, with long pointed and angled wings. Gannets form dense colonies for breeding, with the nests so closely packed that there is only a small space between neighbours. They fly well out over the sea, making spectacular plunge dives from high up, to catch fish. Special air-sacs under the skin cushion the birds against the impact with the water when they dive at great speed.

The Atlantic puffin, seen at sunset *below* and against the moon *right,* is a stubby large-headed bird of the auk family, notable for its highly coloured, compressed bill. This is the breeding dress; the bill is enlarged by the growth of a sheath striped in red, blue-grey and yellow. Puffins fish offshore.

Penguins

There are 17 species of penguin all confined to the southern hemisphere and, except the Galapagos penguins, all in the southern half of it. They are the most specialized of aquatic birds. Their plumage is a distinctive black and white, like that of the Magellanic penguin *left*, any colour being on the head. The strong legs bear webbed feet. The wings are converted to flippers and the body is covered with numerous densely packed small feathers which form a waterproof layer.

Thoroughly at home in water *below*, penguins on land waddle with an upright stance, to all appearances awkwardly. In the water they can swim at 10 knots for leisurely cruising and up to double this figure in short bursts, using the flippers as other birds use their wings in flying. Indeed, it can be said that penguins fly underwater and do so with as much ease and skill as other birds do in the air.

The rockhopper penguin *right* nests on islands all around the Subantarctic. Most penguins are inoffensive. The rockhopper is the most aggressive, ready to jump up at a person and hold on furiously.

Sea Mammals

There are two kinds of mammals that live permanently in the sea. One is the sea cows, which include the dugong and the manatees. The other is the sea otter.

The dugong *left* is found in the Indian Ocean, and eastwards to Western Australia and the Philippines. The several species of manatee *below* live on the coasts and in estuaries on both sides of the tropical Atlantic, and even ascend well up rivers.

Sea cows are descended from similar ancestors to those of elephants. They have lost their hindlegs, their front limbs are changed to paddles and the tail has grown large and bears horizontal flukes, as in whales. Sea cows browse sea grass; manatees eat other aquatic vegetation when they ascend rivers. Manatees have grinding teeth like those of elephants.

The sea otter *right* is not very different from any other otter to look at but it spends its time floating on its back on the kelp beds just offshore, diving when hungry to bring up a mollusk, crab or sea urchin. It also brings up a large flat stone and, using this as an anvil, cracks open its prey by hammering it on the stone.

Seals

Except that they have taken to an aquatic life, seals and their relatives, the sealions and the walrus, are very dog-like. Yet the males are called bulls and the females are called cows, but the babies are known as pups. Their land-living ancestors must have been very nearly related to ancestral dogs.

There are true seals and fur seals, the sealions *centre left and right* being grouped with the latter in classifying them because, like the fur seals they can turn their hind flippers forwards when progressing on land. The hind flippers of true seals are permanently directed backwards. They are effective rudders when the seal is swimming but on land true seals have to hump themselves along like caterpillars. Both the common or harbor seal *top left* and the grey or Atlantic seal *above* are true seals, also called earless seals, whereas the fur seals and sealions have external ears although these are very small. Their teeth have sharp points that help them hold their slippery prey: fishes, octopuses and squids.

Walruses *bottom left* belong to a separate family. Where walruses live the bottom of the sea is mainly gravel. The tusks are used to stir this up and the lips and whiskers are used to sort food and convey it to the mouth. When the stomach of a dead walrus is examined it is usually full of the soft bodies of mollusks, such as cockles. It is believed the walrus must hold the mollusk on the ground with its strong whiskers and suck out the soft body because no shells have ever been found in the stomach.

Dolphins

Since the 1940s, when the first dolphin-arium was built in Florida, literally millions of people have become familiar with these sea mammals. Today, there are a dozen or more dolphinaria, wide-spread over the world, to which visitors flock. Moreover, dolphins are often seen on television. The one vivid impression most people retain is that dolphins are playful. This is true not only for those in captivity but also for those in the wild. Suddenly a school of dolphins swimming near the coast will make a diversion to a sheltered bay and there indulge in a play period, to the delight of anyone on the cliff-top looking down on them.

In this play period can be seen something very similar to our own organized games. Even when travelling for every-day purpose dolphins will suddenly leap from the water, like the Pacific white-sided dolphin *left* and the Pacific bottlenose dolphin *below*. It reminds one of a child who, when walking along a street, suddenly breaks into a hop and a skip.

The dusky dolphin *right* has gone one better, suddenly leaping clear of the surface, and falling back with a great splash, a pastime known as breaching.

Whales

There are two kinds of whales: whalebone whales and toothed whales. The first feed on plankton. They have no teeth. Instead they have plates of baleen or whalebone hanging down from the roof of the mouth. They feed by taking in mouthfuls of water then squeezing it out through the baleen. The plankton is trapped on the inner side of the baleen and swallowed.

When large scale whaling began, first in the Bay of Biscay, later in the Arctic, certain whales were regarded as the 'right' whales to catch. They became known as the right whales. They have been hunted almost to extinction. When the whaling had virtually killed off the whales in the northern hemisphere the whalers went to the South Atlantic and Antarctic. One of the first to be hunted almost to extinction was the southern right whale *left and right*.

The largest of the toothed whales is the sperm whale that has been harried for its oil. This grows to 20 metres in length and feeds mainly on squid. More formidable, although only half the size is the killer whale *below*. It feeds on seals and seabirds, and also on other whales, hunting in packs to attack the large whalebone whales.

Plankton

Plankton is the collective name for plants and animals which drift with the currents, mainly near the surface, in seas, rivers and lakes. Much of it is difficult to see with the naked eye.

Copepods *left* are minute crustaceans which are important elements of the marine plankton. One of the larger forms of plankton, known collectively as krill, are prawn-like *below* and form the food of the large whalebone whales.

The mite *top right* is an uncommon member of the plankton. More common are crab larvae *top far right*, jellyfish *bottom right* and their larvae and fish eggs *bottom far right*. Many fishes lay eggs each of which contains an oil droplet that makes them float at the surface, which they do in their millions.

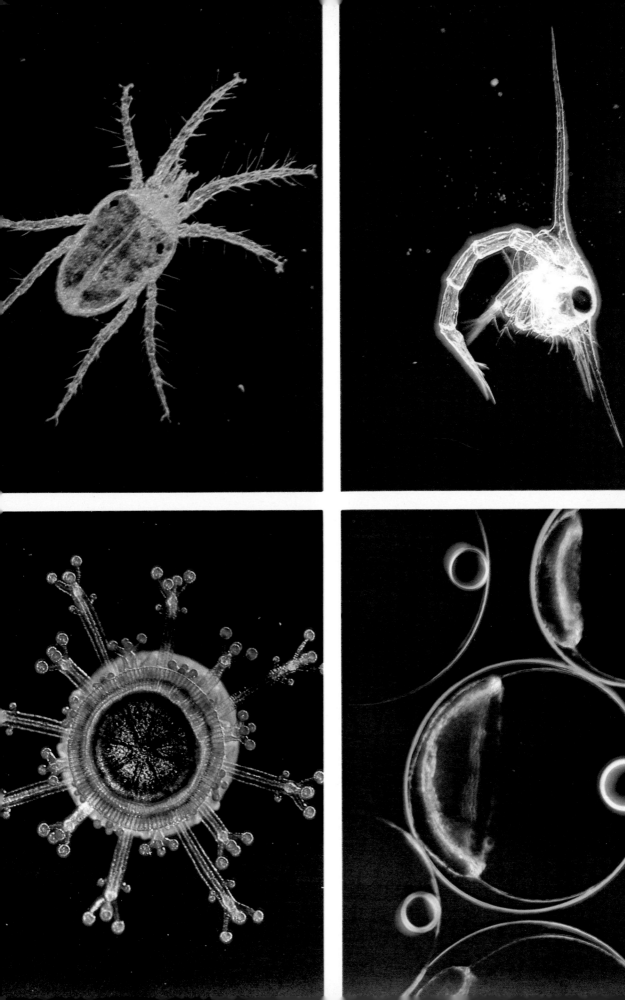

63

INDEX

Page numbers refer to illustrations